MY NONFICTION DECODABLE READERS

Necks, Legs, and Eggs

BY KIM THOMPSON

A Little Honey Book

Crabtree Publishing

crabtreebooks.com

Tips for Teachers and Caregivers

This book supports early readers as they decode words to learn facts and gain knowledge about the world.

Before reading, make sure students understand the sound-spelling correspondences shown below as well as the high-frequency words shown on the next page. Introduce the vocabulary words.

During reading, provide feedback and encouragement as students sound out decodable words by blending individual sounds.

After reading, talk about and write about the topic. Share the information on page 16 to help students learn more.

Letters and Sounds

New:

Sound	Spelling
short e	e
/l/	l

Review:

Sound	Spelling
short a	a
/b/	b
/k/	c, ck, k
/d/	d
/f/	f
/g/	g
/h/	h
short i	i
/m/	m
/n/	n
short o	o
/p/	p
/r/	r
/s/	s
/t/	t
short u	u

Decodable Words

an, and, bed, bend, blends, can, den, dig, egg, fed, get, in, is, it, legs, lets, neck, on, red, snug, tell, ten

High-Frequency Words

New: being, live, long, walk, you, your

Review: a, for, helps, how, the, this, to, water

Vocabulary Words

animal

body

claws

plants

shell

A long neck helps this **animal** live.

It can bend to get **plants**.

animal
plants

Ten **claws** help this animal live.

It can dig a den for a bed.

Being red helps this animal live.

It blends in.

Long legs help this animal live.

It can walk on water.

An egg lets this animal live.

In the **shell**, it is snug and fed.

shell

body

Can you tell how your **body** helps you live?

Build Background Knowledge

Animals need air, food, water, and shelter to survive. Their bodies help them meet all these needs. Specialized body parts help them find food, breathe air, drink water, build safe homes, and more. Look at each animal shown in the book. What body parts help it meet its needs? How does the shape of each body part help it work well? How do your body parts help you live?

MY NONFICTION DECODABLE READERS

Necks, Legs, and Eggs

Written by: Kim Thompson
Designed by: Rhea Magaro
Series Development: James Earley
Educational Consultant: Marie Lemke, M.Ed.

Photographs: All images from Shutterstock

Crabtree Publishing

crabtreebooks.com 800-387-7650

Printed in China/012024/FE20231222

Published in Canada
Crabtree Publishing
616 Welland Ave.
St. Catharines, Ontario
L2M 5V6

Published in the United States
Crabtree Publishing
347 Fifth Ave
Suite 1402-145
New York, NY 10016

Library and Archives Canada Cataloguing in Publication
Available at Library and Archives Canada

Library of Congress Cataloging-in-Publication Data
Available at the Library of Congress

Hardcover: 978-1-0398-4433-9
Paperback: 978-1-0398-4514-5
Ebook (pdf): 978-1-0398-4591-6
Epub: 978-1-0398-4661-6
Read-Along: 978-1-0398-4731-6
Audio: 978-1-0398-4801-6